SpringerBriefs in Educational Communications and Technology

Series editors

J. Michael Spector, University of North Texas, Denton, TX, USA
M.J. Bishop, University System of Maryland in Adelphi, MD, USA
Dirk Ifenthaler, University of Mannheim, Mannheim, Germany,
Deakin University, Geelong, Australia

W0079058

More information about this series at http://www.springer.com/series/11821

Maggie Renken • Melanie Peffer
Kathrin Otrel-Cass • Isabelle Girault
Augusto Chiocarriello

Simulations as Scaffolds in Science Education

 Springer

Maggie Renken
Georgia State University
Atlanta, Georgia, USA

Melanie Peffer
Georgia State University
Atlanta, Georgia, USA

Kathrin Otrel-Cass
Aalborg University
Aalborg, Denmark

Isabelle Girault
Grenoble Alpes University
Grenoble, France

Augusto Chiocarriello
Italian National Research Council
Genova, Italy

ISSN 2196-498X   ISSN 2196-4998   (electronic)
SpringerBriefs in Educational Communications and Technology
ISBN 978-3-319-24613-0   ISBN 978-3-319-24615-4   (eBook)
DOI 10.1007/978-3-319-24615-4

Library of Congress Control Number: 2015953842

Springer Cham Heidelberg New York Dordrecht London

Printed on acid-free paper

Springer International Publishing AG Switzerland is part of Springer Science+Business Media (www.springer.com)

Preface

This book is a product of an international collaboration that began during a preconference workshop of a meeting in the French Alps in 2013. The workshop aimed to develop collaborations for synthesizing the literature regarding inquiry-based science education, scaffolding, and computer simulations through an interdisciplinary perspective. The authors, Drs. Augusto Chiocaricello, Isabelle Girault, Kathrin Otrel-Cass, and Maggie Renken, worked as a small group within a much larger group that came together to discuss general pathways and directions related to technology-enhanced learning (TEL). A postdoctoral research associate working with Dr. Renken, Dr. Melanie Peffer, joined the production of this book project later, after being introduced to the idea and enthusiastic about the nature of the topic. Each author approached this subject from a slightly different background and with experiences relating to their professional pathway; however, all of the authors share a common expertise in science education. We started our discussions after exchanging our ideas in France and then continued writing to sustain our collaboration and in an effort to produce initially an article that later developed into several chapters of a book. We decided that our contribution was not necessarily in rewriting the topic of simulations as scaffolds with new research but rather to bring together, summarize, and expand concepts and how they hang together. Subsequently, chapters were planned, researched, and written, only to be reshuffled into the final shape of this book.

We have designed this book to be an accessible and thorough account and as a result expect the chapters that follow will appeal to educators, researchers, and simulation designers for several reasons. We also see this book making a useful contribution to teacher educators and in encouraging their students to critically examine what the nature of simulations is and how they can be put to good use for science teaching. The book provides a clear overview of the connections between three established fields of study relevant to educators: scaffolding, inquiry- and problem-based approaches to science education, and computer simulations in the classroom. Rather than provide a general overview for integrating simulations in the classroom, we specifically target science education. Unlike most prior work, we have also aimed throughout the chapters to remove the common conflation of scaffolds within simulations and simulations *as* scaffolds—emphasizing simulations as supports for science education. In addition to emphasizing simulations as supports, we address pedagogical considerations and provide explicit instructional recommendations. We expect a careful outline of pedagogical considerations will be informative for simulation designers as they create simulations to support science learning. Finally, by providing a clear outline and discussion of the types of knowledge supported by simulations and instructional recommendations, we hope the book is informative for researchers interested in assessing authentic outcomes, such as students' modeling processes, related to simulations in the classroom.

We hope readers find this book well suited to serve as supplemental reading in undergraduate and graduate coursework and as a principal text for professional development, especially related to integrating technology in the classroom. Many such professional development resources for technology integration or blended learning are subject neutral. On the contrary, this book provides science-specific considerations for teachers integrating simulations in the science classroom. Whether this book offers adequate guidance and impetus for further research and prompts for critical thinking is not for us to say, but we hope it will.

Atlanta, GA, USA   Maggie Renken
Atlanta, GA, USA   Melanie Peffer
Aalborg, Denmark   Kathrin Otrel-Cass
Grenoble, France   Isabelle Girault
Genova, Italy   Augusto Chiocarriello

Contents

The original version of the editor affiliation has been revised. An erratum can be found at
DOI 10.1007/978-3-319-24615-4_7

Chapter 1
An Introduction to Simulations as Scaffolds in Science Education

Maggie Renken, Kathrin Otrel-Cass, Melanie Peffer, Isabelle Girault, and Augusto Chioccariello

The topics of computer simulations, scaffolding, and science education have each been highly studied. In notable recent cases, the connections across these areas of focus have been addressed. For instance, Honey and Hilton's (2011) report, *Learning Science Through Computer Games and Simulations*, provides an exhaustive review of simulations as tools for learning science. Also in a seminal paper, Quintana et al. (2004) provide a framework for the ways in which students' inquiry learning in science can be scaffolded via software. However, a decade after this framework was presented, significant challenges continue to vex those aiming to support student learning through computer-based scaffolds (Rienties et al. 2012). For instance, researchers and educators alike continue to struggle with how best to identify and respond to complex individual learner needs, such as those related to balancing needs for support and freedom. In response, calls like those by Honey and Hilton (2011) charge researchers, developers, and educators to consider how and to what extent individually different science learners in diverse classroom settings benefit from simulations in science education. We respond to this charge here with a synthesized overview of the literature and key issues related to simulations as scaffolds within science education settings. We choose to focus particularly on inquiry- and problem-based science education because, as we will outline in further detail, such open-ended exploratory learning environments align with recent trends in international science education standards, are often challenging to facilitate, and, subsequently, are likely to benefit from appropriate scaffolds.

Broadly, traditional consideration of the intersection of computer simulations and scaffolds has come in two predominant forms: internal scaffolds and external scaffolds to support learning *within* the simulation context. Internal scaffolds are those embedded in the simulation, while external scaffolds are those provided, from

---

The original version of this chapter was revised. The erratum to this chapter is available at: DOI 10.1007/978-3-319-24615-4_7

© AECT 2016

M. Renken et al., *Simulations as Scaffolds in Science Education*, SpringerBriefs in Educational Communications and Technology, DOI 10.1007/978-3-319-24615-4_1

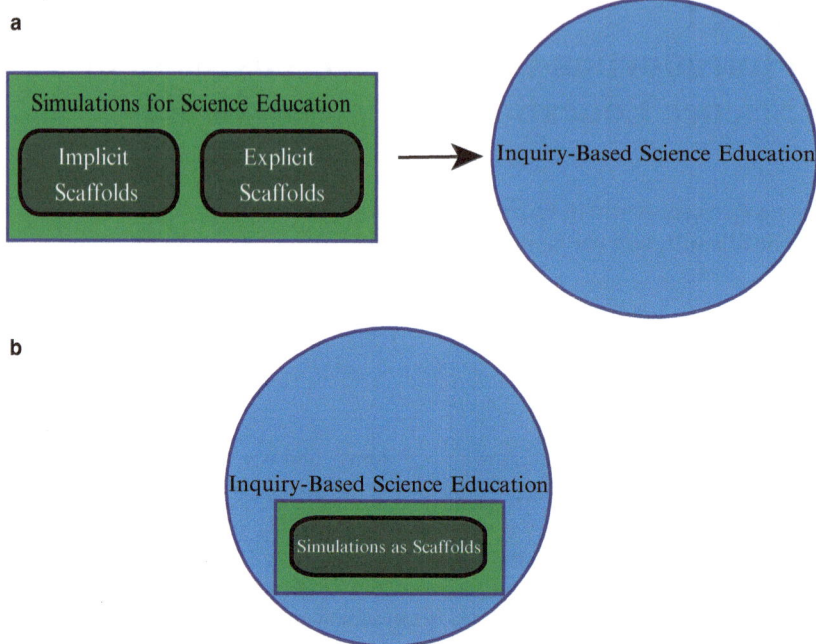

Fig. 1.1 (**a**) Traditional consideration of the intersection of simulations and scaffolds has focused on internal and external scaffolds that support learning within the simulation context. (**b**) The perspective presented here is of computer simulations as scaffolds embedded within broader science education curricula

perhaps a teacher, to support student use of a simulation (Honey and Hilton 2011; see discussion of value-added approach in Mayer 2014 for examples of internal scaffolds). The perspective we offer here differs by considering simulations as the scaffolds to be embedded within broader science education curricula (Fig. 1.1). We make this distinction for two important reasons. First, embedding simulations as scaffolds within science education curricula is more reflective of authentic science practice. Second, embedding simulations as scaffolds within curricula may have benefits that are specific to students' science learning outcomes, especially those related to students' motivation and interest and to their understanding of authentic science practices and disciplinary core knowledge.

While computer simulations have been employed in classrooms as valuable tools for promoting varied science learning outcomes, some prior research suggests these outcomes do not differ when students use simulations versus physical, hands-on experiments (Clark et al. 2009; de Jong et al. 2013; Renken and Nunez 2013). Instead it is increasingly evident that students learn best with a combination of physical and virtual manipulatives (de Jong et al. 2013; Kontra et al. 2015). Our primary aim is to describe how simulations may best be harnessed as scaffolds to support science education objectives. With this approach, we address Pea's (2004) challenge to researchers to ask how scaffolding can be related to specific teaching and learning goals. In what follows, we provide an overview of discussions relevant to the topic of simulations as scaffolds to support science education. In Chap. 2, we define computer simulations for our purposes. In Chap. 3, we elaborate on our definition of computer simulations by drawing distinctions between simulations and other educational technologies, including static animations, serious games, and virtual worlds. We then briefly describe the motivations, objectives, and challenges associated with inquiry-based science education and a problem-based learning approach to science education in Chap. 4. In Chap. 5, we describe scaffolding and identify three categories of science learning in need of scaffolding. Based on this overview, in Chap. 6, we outline key issues for addressing the grand challenges posed to educators, developers, and researchers interested in embedding simulations within science education while keeping learning principles in mind. We explore the use of computer simulations as instructional scaffolds that provide strategies and support when students are faced with the need to acquire new skills or knowledge. A goal of this brief is to provide insight into what research has reported on navigating the complex process of inquiry- and problem-based approaches to science education and how computer simulations may best contribute to that task.

References

Clark, D. B., Nelson, B., Sengupta, P., & D'Angelo, C. (2009). *Rethinking science learning through digital games and simulations: Genres, examples, and evidence.* Paper commissioned for the National Research Council workshop on gaming and simulations.

de Jong, T., Linn, M. C., & Zacharia, Z. C. (2013). Physical and virtual laboratories in science and engineering education. *Science, 340*(6130), 305–308. doi:10.1126/science.1230579.

Honey, M., & Hilton, M. (Eds.). (2011). *Learning science through computer games and simulations.* Committee on Science Learning: Computer Games, Simulations, and Education. Board on Science Education, Division of Behavioral and Social Sciences and Education. Washington, DC: The National Academies Press.

Kontra, C., Lyons, D. J., Fischer, S. M., & Beilock, S. L. (2015). Physical experience enhances science learning. *Psychological Science, 26*(6), 737–749. doi:10.1177/0956797615569355.

Mayer, R. (2014). *Computer games for learning: An evidence-based approach.* Cambridge, MA: MIT Press.

Pea, R. D. (2004). The social and technological dimensions of scaffolding and related theoretical concepts for learning, education, and human activity. *The Journal of the Learning Sciences, 13*(3), 423–451. Stable URL: http://www.jstor.org/stable/1466943.

Quintana, C., Reiser, B. J., Davis, E. A., Krajcik, J., Fretz, E., Duncan, R. G., et al. (2004). A scaffolding design framework for software to support science inquiry. *The Journal of the Learning Sciences, 13*(3), 337–386. doi:10.1207/s15327809jls1303_4.

Renken, M. D., & Nunez, N. (2013). Computer simulations and clear observations do not guarantee conceptual understanding. *Learning and Instruction, 23*, 10–23. doi:10.1016/j.learninstruc.2012.08.006.

Rienties, B., Giesbers, B., Tempelaar, D. T., Lygo-Baker, S., Segers, M., & Gijselaers, W. H. (2012). The role of scaffolding and motivation in CSCL. *Computers & Education, 59*(3), 893–906. doi:10.1016/j.compedu.2012.04.010.

Chapter 2
Computer Simulations on a Multidimensional Continuum: A Definition and Examples

Isabelle Girault, Melanie Peffer, Augusto Chioccariello,
Maggie Renken, and Kathrin Otrel-Cass

Although simulations were initially developed to solve "unsolvable" scientific equations, they are now used for a variety of practical reasons both in research and in classrooms (Greca et al. 2014). Starting in the 1950s, simulations were used to examine a simplified representation of an actual system and consequently allowed for the first virtual experiments. At its core, a simulation allows the user to manipulate the variables of a model in order to observe the dynamic evolution of the system's state, visualized on a screen. With the advent of direct manipulation and sophisticated graphical user interfaces, simulations have become highly interactive and accessible not only for researchers but within the classroom as well. Computer simulations for the classroom have followed the rapid growth of more general computer-based applications. In PhET chemistry simulations, student users can manipulate parameters within the simulation (e.g., changing the concentration of solute within a solution) and view feedback in real time, such as observing a color change (Moore et al. 2014). Furthermore, students can view multiple representations of a single entity. In the case of the Models of the Hydrogen Atom PhET simulation, students can manipulate the simulation to view six different atomic models (Clark and Chamberlain 2014; see Fig. 4).

Simulations exist with other educational technologies, such as static animations and serious games along a multidimensional continuum (Fig. 2.1) (c.f., D'Angelo et al. 2014). This continuum is multidimensional because static animations, simulations, and serious games relate to one another in terms of their shared use of multiple features. For instance, simulations relate to static animations in terms of their shared use of visualization and to serious games in terms of their shared use of models and entertainment features. Although multiple features may be shared across forms of educational technology, the scope of these features differs. Because

---

The original version of this chapter was revised. The erratum to this chapter is available at: DOI 10.1007/978-3-319-24615-4_7

© AECT 2016
M. Renken et al., *Simulations as Scaffolds in Science Education*,
SpringerBriefs in Educational Communications and Technology,
DOI 10.1007/978-3-319-24615-4_2

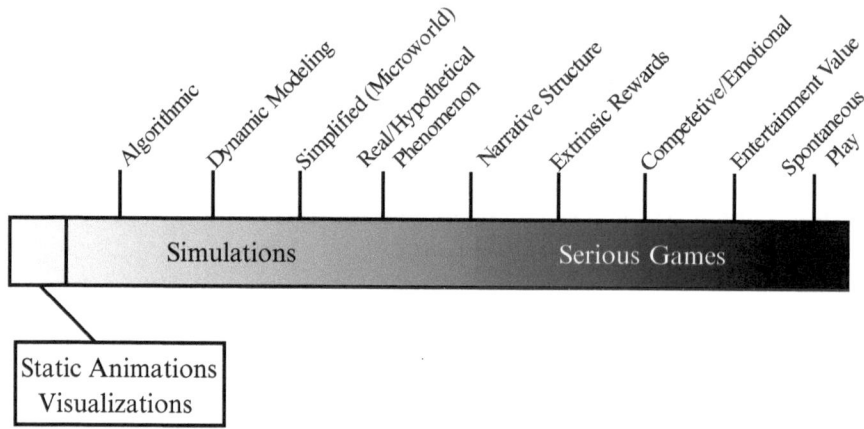

Fig. 2.1 Continuum of educational technologies relevant to the science classroom

Table 2.1 Definitions of computer simulations

Citation	Definition of computer simulation
Winsberg (2015)	"A program that is run on a computer and that uses step-by-step methods to explore the approximate behavior of a mathematical model. Usually this is a model of a real-world system (although the system in question might be an imaginary or hypothetical one). ... When run on a computer, the algorithm produces a numerical picture of the evolution of the system's state, as it is conceptualized in the model" (Section 1.1)
Honey and Hilton (2011)	"Computational models of real or hypothesized situations or natural phenomena that allow users to explore the implications of manipulating or modifying parameters within them. ... Simulations allow users to observe and interact with representations of processes that would otherwise be invisible" (p. 9)
Hirumi (2010)	"A dynamic and testable representation of an activity, behavior, process, situation, or combination thereof whose parameters can be manipulated and studied. Simulations used to acquire new knowledge…are representative artificial worlds in which research questions from a range of fields and disciplines can be tested and analyzed" (p. 88)
D'Angelo et al. (2014)	"A tool used to explore a real-world or hypothetical phenomenon or system by approximating the behavior of the phenomenon or operation of the system. ... Simulations…must be constructed with an underlying model that is based on some real-world behavior or natural/scientific phenomenon (such as models of the ecosystem or simulated animal dissections). …The simulation includes some interactivity on the part of the user, centered usually on inputs and outputs or more generally, the ability to set parameters for modeling the phenomenon or system" (p. 2)

these technology-based tools lie on a continuum with regard to multiple features, many different definitions for simulations exist, and the lines between simulations and other technologies can be diaphanous. Table 2.1 lists several comprehensive definitions of simulations present in recent literature. These definitions overlap in

Table 2.2 Examples of simulations in which the key features of computer simulations promote the exploring of ideas, manipulating parameters, observing events, and testing questions

	Exploring ideas	Solve problem/ test questions	Manipulate parameters	Observe/study otherwise invisible features
Algorithmic			PhET	
Dynamic modeling	Dance of the planets	BioLogica	PhET Dance of the planets	PhET
Simplified (microworld)		PhET	PhET NetLogo	PhET NetLogo Investigations in electromagnetism
Real/hypothetical phenomenon	ChemSense OsmoBeaker	BioLogica Thinker Tools OsmoBeaker	Thinker Tools OsmoBeaker	ChemSense OsmoBeaker NetLogo Investigations in electromagnetism

their descriptions of computer simulation features. Based on these definitions, and to distinguish simulations from other forms of educational technologies relevant to science learning (e.g., serious games), we define simulations relevant to scaffolding in science education as *algorithmic, dynamic, often simplified models of real-world or hypothetical phenomenon that contain features that not only allow but promote the exploration of ideas, manipulation of parameters, observation of events, and testing of questions.* In the remainder of this section, we elaborate on each component of our definition of simulations with examples of simulations used for science education. Table 2.2 is organized around such examples of simulations that demonstrate how certain features allow certain outcomes.

Simulations Are Algorithmic

All simulations contain some underlying rule structure or algorithm. What algorithms are present will dictate how the simulation functions. For example, in BioLogica's Dragon Genetics simulation, a predetermined rule structure dictates which genotype will give rise to a particular feature (Buckley et al. 2004). If a student changes genes related to color, only the dragons' color, not their wing type, will be impacted. The algorithmic rule-based structure of simulations is most apparent in quantitative simulations where the user is constrained by natural constraints (e.g., known electrostatic forces between molecules) to model phenomena such as molecular dynamics. (See *Simulation models may be either conceptual or computational in nature.*) Although this underlying rule structure can be useful for teaching students the constraints of the simulated phenomenon, a simulation is only as good as the rule structure on which it is based. If the algorithms driving a rule structure are incomplete or inaccurate, the simulation may lead students to generate misconceptions (Hirumi 2010).

Simulations Are Dynamic

Unlike static animations, simulations allow for interactivity in which selected parameters can be changed, and the impact of such changes can be observed and at times analyzed. The emphasis here lies in simulations as a tool for exploring ideas. Such "explorations" involve varying degrees of interaction with the simulation. For example, in some simulations like BioLogica's Dragon Genetics (Fig. 2.2), the user manipulates parameters from "outside" the context of the simulation—in this case the genes the dragon will inherit. As the user manipulates these genes, the appearance of the dragon changes in real time. Simulations, such as the Peppered Moth simulation (http://peppermoths.weebly.com/), place the user in the role of an animal or avatar interacting with the simulation environment. In the Peppered Moth simulation, the user is tasked as a bird and told to eat as many moths as possible. Since some moths blend in with the environment better than others, they are easier to see and consequently eat. The simulation gives the student real-time feedback as to how many black and white moths are left in the forest at any given time. Alternatively, some simulations such as the Science Classroom Inquiry (SCI) simulations allow for open-ended exploration where students are given autonomy to complete the simulation as they wish (Peffer et al. 2015).

Fig. 2.2 The BioLogica simulation, *Dragon Genetics*, allows students to explore the relationship between genotype and phenotype. *Screenshot image* used with permission from the Concord Consortium (www.concord.org)

Simulations Are Often Simplified

Teachers are often required to reduce complexity and simplify ideas that are being studied. A lack of simplified representations can be an obstacle to supporting student learning. Being able to research without the constraining variables of time and space makes the simulation a particularly useful resource in the classroom (Urban-Woldron 2009). A good example is Bohr's model of the atom represented as a nucleus surrounded by electrons at a distance (Fig. 2.3). This simplification is commonly used to help students conceive of what an atom might look like and to explain abstract ideas such as line spectra. This simplified model is used at the secondary school level, while the modern quantum mechanical model of the atom is left to advanced courses (Osborne 2011).

Simulations may also be part of microworlds. Microworlds are similar to simulations in that they model real-world phenomena on a simpler scale but with a more flexible rule structure allowing for student interaction to be more open ended. Rieber (2005) suggests that microworlds are more akin to programming languages because the user manipulates a language like LOGO to modify the microworld. Consequently, students are not limited to what the designer intended but can create their own models and explore their own understanding (Rieber 2005). diSessa (2000) describes microworlds as containing "an easy-to-understand set of operations that students can use to engage tasks of value to them, and in doing so, they come to understand powerful underlying principles" (p. 47). Papert (1980) defines

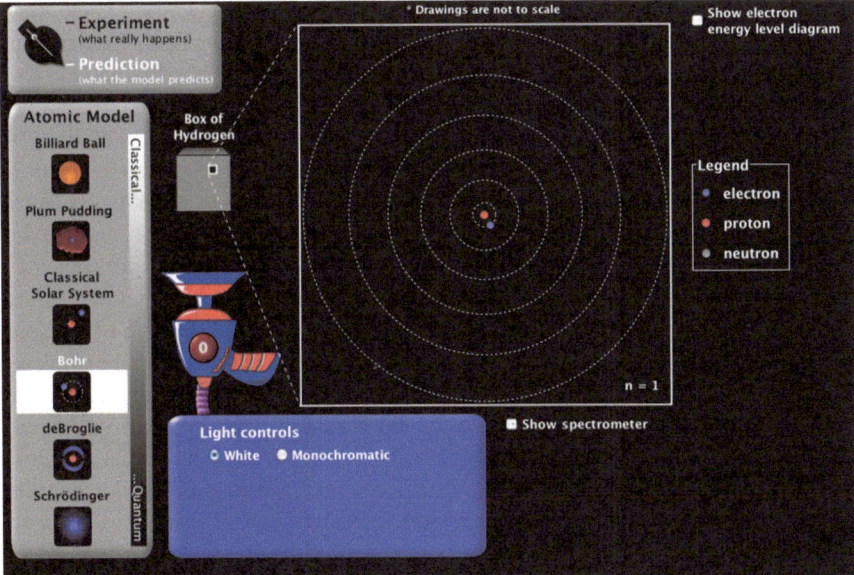

Fig. 2.3 Models of the hydrogen atom: Bohr's model, PhET interactive simulations, University of Colorado Boulder, http://phet.colorado.edu

microworlds as a "…subset of reality or a constructed reality whose structure matches that of a given cognitive mechanism so as to provide an environment where the latter can operate effectively. The concept leads to the project of inventing microworlds so structured as to allow a human learner to exercise particular powerful ideas or intellectual skills" (p. 204). Therefore, like simulations, microworlds can be simplified representations of the real world but are distinct from simulations in that the user has more freedom to manipulate the microworld to test their own models.

Simulations Are Models

Simulations are used as models in a variety of fashions. Simulations may model real-world or hypothetical phenomenon in either a conceptual or computational fashion. Simulations may also be used as science tools by scientists to test and generate new knowledge or to provide models for students to test questions. Simulations also are differentiated in terms of learning that occurs by exploring a model created by someone else and learning from constructing one's own model.

(A) Simulations Model Real-World or Hypothetical Phenomenon

From the first applications in nuclear physics and meteorology, simulations have become a tool routinely used to model and understand a wide range of natural and hypothetical phenomena. Examples of simulations used to represent natural phenomena include the NetLogo Investigations in Electromagnetism (NIELS) and Dragon Genetics simulation from BioLogica (Sengupta and Wilensky 2009; Buckley et al. 2004, respectively). NIELS simulations allow students to visualize electrons moving within a wire and manipulate those electrons to better understand electric current. In the Dragon Genetics simulations, students explore genetics concepts and can test their own ideas about which genotypes lead to certain phenotypes (Fig. 2.2).

Yet simulations are not limited only to reproduce actual phenomena; they allow users to model hypothetical inquiries as well, such as how the strength of the solar wind might influence the orbit of planets or what kind of sea tides could be observed on an Earth with two moons. Furthermore, as we discussed in the previous section, simulations allow for students to examine simplified examples of real-world phenomena. Simplifying real-world examples and allowing for hypothetical scenarios allow students to overcome challenges related to explanations of phenomena that are at times counterintuitive and not supported by everyday experience, at least not until one has learned to "read" that experience in very specific ways (Millar 2004). Thus, for the learner, a simulation can be used as a cognitive tool to look at phenomena through the "lenses" of established knowledge—starting, stopping, examining, reexamining, or restarting investigations under new conditions. In other words, simulations offer learners the interrogation of situations that are impossible to witness in the same way in reality, such as stopping a ball in mid-flight or rewinding time. Those features allow students to explore the mechanisms underlying scientific phenomena that contribute to their growing conceptual understanding (Çepni et al. 2006; Clark et al. 2009; Honey and Hilton 2011).

Simulations that model real-world and hypothetical phenomenon are used not only in educational settings but also as tools for scientists and engineers. When scientists use simulations, they often model complex and sophisticated processes. For example, in the field of molecular biology, molecular dynamics simulations use knowledge of the atomic structures of various molecules and the underlying laws of physics to approximately model such biological processes such as membrane transport, ligand binding, conformational change, and protein folding (Dror et al. 2012). Simulations that model molecular dynamics may be useful in the future for facilitating the design of novel pharmacologic targets and also for identifying previously unknown structural aspects of certain biomolecules (Dror et al. 2012). Simulations are also used in the generation of new knowledge. For example, simulations can be based on a rudimentary theory and used to test hypotheses about how that theory governs various natural phenomena (Greca et al. 2014). As such, computer simulations allow both novices and researchers alike to ask and test questions without risking life and property (Harrison and Treagust 2000).

Although some simulations are used for both research and education, simulations for science education have different goals (e.g., helping the student develop an understanding of a certain phenomena), require guidance to reach maximal efficacy, and tend to be either virtual laboratories or replications of natural phenomena that the student interacts with (Greca et al. 2014). While the scientist or expert user seeks to create new knowledge through the simulation, the student or novice typically seeks to understand the major elements of a body of preestablished, consensually agreed-upon knowledge (Osborne 2011).

(B) Simulation Models May Be Either Conceptual or Computational in Nature

Simulations invite users to explore ideas by providing models to test questions or solve problems (Otrel-Cass 2001). As described in the preceding section, simulations contain an underlying rule structure or algorithm that is used to model a phenomenon. The flexibility of such rule structures varies depending on if a simulation is modeling something of a conceptual or computational nature. Simulations that contain conceptual models tend to be qualitative, abstract, and allow the user to build on current understanding through exploration. Conceptual simulations include simulations that model abstract concepts like authentic science practices. For example, Science Classroom Inquiry (SCI) simulations are designed to allow students to engage in authentic inquiry within the confines of a typical classroom. Students are tasked as researchers trying to solve a research question, such as determining why abnormally large numbers of a certain species are dying in the wild. As students complete their investigations, the simulation provides scaffolding. Rather than simulating a virtual laboratory, SCI simulates the thinking process necessary for successful authentic inquiry, promoting conceptual understanding of authentic science practices *and* content (Peffer et al. 2015). In contrast, computational, or quantitative, simulations follow a detailed and rigid structure, often times based in natural rules (e.g., gravitational pull). Computational modeling includes tools such as GROMACS (Lindahl et al. 2001), which allows users to model molecular dynamics. GROMACS is used by both researchers and within the undergraduate classroom (Elmore et al. 2010).

(C) Simulations Provide an Opportunity for Students to Build Their Own Models

As discussed in the preceding section, simulations function as models in a wide variety of contexts. They may serve to model real-world phenomena in a constrained, quantitative manner. In fact, when we discuss students' "modeling" phenomena via simulations, we most often are referring to learning from models created by someone else (e.g., the simulation designer). Alternatively, simulations may allow students to apply and test models they have created on their own. As an example of such "metamodeling," the Modeling and Inquiry Learning Application, or MILA, technology allows users to build conceptual models and then test those models using an executable simulation (Joyner et al. 2014). Students are first tasked with developing a conceptual model within the software to explain an observed ecological phenomenon such as why fish are dying within a lake. After building the model, the software allows students to then simulate their model using the NetLogo simulation platform. After viewing the simulation, students can then go back and revise their original conceptual models. Consequently, the integration of conceptual model building with an executable simulation allows for students to test and revise their conceptual understanding.

With a definition of simulations in the context of science education set, in the following section of this brief, we clarify distinctions between simulations and other prominent educational technologies. In other words, we have established what a simulation is, next we establish what a simulation is *not*.

References

Buckley, B. C., Gobert, J. D., Kindfield, A. C. H., Horwitz, P., Tinker, R. F., Gerlits, B., et al. (2004). Model-based teaching and learning with BioLogica: What do they learn? How do they learn? How do we know? *Journal of Science Education and Technology, 13*(1), 23–41. doi:10.1023/B:JOST.0000019636.06814.e3.

Çepni, S., Taş, E., & Köse, S. (2006). The effects of computer-assisted material on students' cognitive levels, misconceptions and attitudes towards science. *Computers & Education, 46*(2), 192–205. doi:10.1016/j.compedu.2004.07.008.

Clark, T. M., & Chamberlain, J. M. (2014). Use of a PhET interactive simulation in general chemistry laboratory: Models of the hydrogen atom. *Journal of Chemical Education, 91*, 1198–1202. doi:10.1021/ed400454p.

Clark, D. B., Nelson, B., Sengupta, P., & D'Angelo, C. (2009). *Rethinking science learning through digital games and simulations: Genres, examples, and evidence.* Paper commissioned for the National Research Council workshop on gaming and simulations.

D'Angelo, C., Rutstein, D., Harris, C., Bernard, R., Borokhovski, E., & Haertel, G. (2014). *Simulations for STEM learning: Systematic review and meta-analysis (executive summary).* Menlo Park, CA: SRI International.

diSessa, A. A. (2000). *Changing minds: Computers, learning, and literacy.* Cambridge, MA: MIT Press.

Dror, R. O., Dirks, R. M., Grossman, J. P., Xu, H., & Shaw, D. E. (2012). Biomolecular simulation: A computational microscope for molecular biology. *Annual Review of Biophysics, 41*(1), 429–452. doi:10.1146/annurev-biophys-042910-155245.

Elmore, D. E., Guayasamin, R. C., & Kieffer, M. E. (2010). A series of molecular dynamics and homology modeling computer labs for an undergraduate molecular modeling course. *Biochemistry and Molecular Biology Education, 38*(4), 216–223. doi:10.1002/bmb.20396.

Greca, I. M., Seoane, E., & Arriassecq, I. (2014). Epistemological issues concerning computer simulations in science and their implications for science education. *Science and Education, 23*, 897–921. doi:10.1007/s11191-013-9673-7.

Harrison, A. G., & Treagust, D. F. (2000). Learning about atoms, molecules, and chemical bonds: A case study of multiple-model use in grade 11 chemistry. *Science Education, 84*(3), 352–381. doi:10.1002/(SICI)1098-237X(200005)84:3<352::AID-SCE3>3.0.CO;2-J.

Hirumi, A. (Ed.). (2010). *Playing games in school: Video games and simulations for primary and secondary education.* International Society for Technology in Education.

Honey, M., & Hilton, M. (Eds.). (2011). *Learning science through computer games and simulations.* Committee on Science Learning: Computer Games, Simulations, and Education. Board on Science Education, Division of Behavioral and Social Sciences and Education. Washington, DC: The National Academies Press.

Joyner, D. A., Goel, A. K., & Papin, N. M. (2014). MILA-S: Generation of agent-based simulations from conceptual models of complex systems. In *Proceedings of the 19th international conference on intelligent user interfaces* (pp. 289–298). ACM.

Lindahl, E., Hess, B., & van der Spoel, D. (2001). GROMACS 3.0: A package for molecular simulation and trajectory analysis. *Journal of Molecular Modeling, 7*, 306–317. doi:10.1007/s008940100045.

Millar, R. H. (2004). *The role of practical work in the teaching and learning of science.* Paper prepared for the Committee: High School Science Laboratories: Role and Vision, National Academy of Sciences, Washington, DC.

Moore, E. B., Chamberlain, J. M., Parson, R., & Perkins, K. K. (2014). PhET interactive simulations: Transformative tools for teaching chemistry. *Journal of Chemical Education, 91*, 1191–1197. doi:10.1021/ed4005084.

Osborne, J. (2011). Science teaching methods: A rationale for practices. *School Science Review, 93*, 93–103.

Otrel-Cass, K. (2001). *Earth science in New Zealand Science Centres—Learning aspects through a simulation based experience.* Unpublished Doctoral Dissertation, University of Waikato, Hamilton, New Zealand.

Papert, S. (1980). Computer-based microworlds as incubators for powerful ideas. In R. Taylor (Ed.), *The computer in the school: Tutor, tool, tutee* (pp. 203–210). New York: Teacher's College Press.

Peffer, M. E., Beckler, M. L., Schunn, C., Renken, M., & Revak, A. (2015). Science Classroom Inquiry (SCI) simulations: A novel method to scaffold science learning. *PLoS ONE, 10*(3), e0120638. doi:10.1371/journal.pone.0120638.

Rieber, L. P. (2005). Multimedia learning in games, simulations, and micorworlds. In R. E. Mayer (Ed.), *The Cambridge Handbook of Multimedia Learning* (pp. 549–567). New York: Cambridge University Press.

Sengupta, P., & Wilensky, U. (2009). Learning electricity with NIELS: Thinking with electrons and thinking in levels. *International Journal of Computers for Mathematical Learning, 14*(1), 21–50. doi:10.1007/s10758-009-9144-z.

Urban-Woldron, H. (2009). Interactive simulations for the effective learning of physics. *Journal of Computers in Mathematics and Science Teaching, 28*(2), 163–176.

Winsberg, E. (2015). Computer simulations in science. In E. N. Zalta (Ed.), *The Stanford encyclopedia of philosophy* (Summer 2015 Edition). Stable URL: http://plato.stanford.edu/archives/sum2015/entries/simulations-science/.

Chapter 3
Distinctions Between Computer Simulations and Other Technologies for Science Education

**Melanie Peffer, Maggie Renken, Isabelle Girault,
Augusto Chioccariello, and Kathrin Otrel-Cass**

Simulations exist along a continuum with other educational technologies (D'Angelo et al. 2014). To further elaborate on our definition of simulations, in this section we describe the distinctions between simulations and three other such technologies: static animations, serious games, and virtual worlds. Rather than provide an exhaustive review of these technologies, we purposefully emphasize distinctions that are relevant to our aim of employing simulations as tools for navigating complex challenges in science education.

(A) Simulations Are Not Static Animations

Simulations are not to be confused with static animations or visualizations. Static animations and visualizations do not give the person engaged in the process an opportunity to manipulate parameters and thereby modify results. Static manipulations may take the form of describing the parabolic path taken by a ball through the air or the water cycle. These animations would "cycle" through without any option for the user to interact with or manipulate any features. The ability for a user to interact with and manipulate the computerized visualization is an essential distinguishing factor between simulations and static animations.

The original version of this chapter was revised. The erratum to this chapter is available at: DOI 10.1007/978-3-319-24615-4_7

M. Renken et al., *Simulations as Scaffolds in Science Education*,
SpringerBriefs in Educational Communications and Technology,
DOI 10.1007/978-3-319-24615-4_3

(B) Simulations Are Not Serious Games

Piggybacking on the increasing popularity of video and computer games, serious games are a form of computer-based educational technology recently introduced to the science classroom. The user of a game often decides if the goal is for entertainment or for non-entertainment, serious purposes. For example, the popular Nintendo Wii® system offers various sports and fitness games that the user could play for fun, or with the express health goal of losing weight or maintaining fitness (Honey and Hilton 2011). Typically, games designed for science learning are considered *serious* as the users undertake them as part of a classroom activity, rather than as part of extracurricular leisure time. Well-designed serious games that effectively link engaging, interactive activity with science education have the potential to be highly motivating for students and to reach large audiences (Honey and Hilton 2011; Mayo 2009).

On the continuum of computer-based educational technologies, simulations differ from serious games. Serious games typically have entertainment aims, follow a narrative structure, and promote engagement via extrinsic rewards, such as points. Serious games also typically have a quantifiable end goal, toward which the user receives feedback on his or her progress in the form of a reward, such as points. We distinguish serious games from simulations using two main criteria (Clark et al. 2009; Honey and Hilton 2011). Generalizing across these two criteria, we conclude motivational scaffolds are not embedded within simulations to the extent they are embedded within serious games.

1. Serious Games for Science Learning Have a Clear Goal Structure that Is Often Reward Based

Serious games often follow a narrative format, progressing through the accomplishment of goals. For example, in SURGE, students play as a female astronaut whose mission is to save a species from destruction. To accomplish this mission, students are tasked with completing a series of challenges. To move from one challenge to the next, the student learns and uses various principles of Newtonian mechanics (Clark et al. 2011). In Quest Atlantis, students are engaged in the virtual world of Atlantis, a multiuser virtual environment (MUVE). Within this MUVE, students are assigned goals or quests to complete. These quests can take a variety of forms depending on the overall goals of the classroom. Furthermore, students can earn points for completing quests and certain in-game "bonuses," such as earning the ability to fly (Barab et al. 2005). In the case of EteRNA, an Internet-based citizen science game available through PBS NOVA (http://www.pbs.org/wgbh/nova/labs/about-rna-lab/), after players complete a certain number of puzzles related to RNA folding and structure, they can design RNA structures that may be used for actual empirical testing in wet laboratory settings. Player-generated structures were better than previously used algorithms indicating that serious gameplay can lead to advances in the sciences (Lee et al. 2014). In contrast, simulations do not have explicit, rewarded goals. Instead, the goal is often the modeling or exploration task itself. While intrinsic, progressive goals may exist on a trajectory toward a larger learner goal of modeling some phenomenon, these are different than the explicit goal structures that define serious games.

2. Serious Games for Science Learning Have Both Learning and Entertainment Objectives

Serious games typically promote learning within a "fun" context. We use "fun" here to describe contexts that are generally considered as entertaining or enjoyable. For example, games that promote competition between players are often categorized as entertaining. Alternatively, allowing for users to approach games that have increasing challenges or goals can be also considered entertaining.

Promoting an engaging, motivational environment is especially crucial when utilizing serious gaming in informal learning environments, such as a zoo or museum. For example, educators at the Minnesota Zoo developed WolfQuest (www.wolfquest.org), a serious game in which users act as wolves in Yellowstone National Park. Gameplay is directed at performing normal wolf behaviors, like hunting, in order to survive (i.e., goal) (Honey and Hilton 2011). Although the game does not explicitly instruct players about wolves, participants report increased emotional investment in wolves, a desire to seek more information about wolves, and increased understanding of wolf biology as a result of playing the serious game (Goldman et al. 2009). Although simulations are intrinsically entertaining as well, unlike serious games, they are not purposefully designed to be entertaining. Furthermore, simulations designed to serve as scientific tools capable of modeling real or hypothetical phenomena may not have any entertainment value.

(C) Simulations Are Not Virtual Worlds

Perhaps the clearest distinction between virtual worlds and simulations is that virtual worlds can serve as environments for housing simulations and not vice versa. A virtual world is created either through graphical or text representation of a real or imaginary space. Virtual worlds are inherently flexible in nature and facilitate multiple learning objectives and goals while giving total student autonomy to explore the world. For example, in Whyville (www.whyville.net), students independently explore a virtual city to learn about various career paths (Kafai 2010). When integrating simulations into the larger context of the science classroom, simulations embedded in virtual worlds may be a useful pedagogical technique.

References

Barab, S., Thomas, M., Dodge, T., Carteaux, R., & Tuzun, H. (2005). Making learning fun: Quest Atlantis, a game without guns. *Educational Technology Research and Development, 53*(1), 86–107. doi:10.1007/BF02504859.

Clark, D. B., Nelson, B., Sengupta, P., & D'Angelo, C. (2009). *Rethinking science learning through digital games and simulations: Genres, examples, and evidence.* Paper commissioned for the National Research Council workshop on gaming and simulations.

Clark, D. B., Nelson, B. C., Chang, H. Y., Martinez-Garza, M., Slack, K., & D'Angelo, C. M. (2011). Exploring Newtonian mechanics in a conceptually-integrated digital game: Comparison of learning and affective outcomes for students in Taiwan and the United States. *Computers and Education, 57*(3), 2178–2195. doi:10.1016/j.compedu.2011.05.007.

D'Angelo, C., Rutstein, D., Harris, C., Bernard, R., Borokhovski, E., & Haertel, G. (2014). *Simulations for STEM learning: Systematic review and meta-analysis (executive summary).* Menlo Park, CA: SRI International.

Goldman, K. H., Koepfler, J., & Yocco, V. (2009). *WolfQuest summative evaluation: Full summative report.* Institute for Learning Innovation. http://www.informalscience.org/images/evaluation/WQ_Full_Summative_Report.pdf. Retrieved 2015.

Honey, M., & Hilton, M. (Eds.). (2011). *Learning science through computer games and simulations.* Committee on Science Learning: Computer Games, Simulations, and Education. Board on Science Education, Division of Behavioral and Social Sciences and Education. Washington, DC: The National Academies Press.

Kafai, Y. B. (2010). World of Whyville: An introduction to tween virtual life. *Games and Culture, 5*(1), 3–22. doi:10.1177/1555412009351264.

Lee, J., Kladwang, W., Lee, M., Cantu, D., Azizyan, M., Kim, H., et al. (2014). RNA design rules from a massive open laboratory. *Proceedings of the National Academy of Sciences, 111*(6), 2122–2127.

Mayo, M. (2009, January). Video games: A route to large-scale STEM education? *Science,* 79–82. http://www.sciencemag.org/content/323/5910/79.short. Retrieved 2015.

Chapter 4
Inquiry-Based Science Education and Problem-Based Learning: Motivations, Objectives, and Challenges Relevant to Computer Simulations

Kathrin Otrel-Cass, Maggie Renken, Melanie Peffer, Isabelle Girault, and Augusto Chioccariello

School science has been criticized for lacking authenticity and relevance and has been associated with students' lack of interest in science beyond the compulsory years at school (Bolstad and Hipkins 2008; OECD 2006). Fensham (2006) argues that if the problem of declining student interest in and motivation for learning science is to be addressed, science education will need to provide opportunities for students where they can connect with real-world science and technology issues. Internationally, claims are being made about the potential of inquiry-based learning to address the challenges of relevance for school science (Aikenhead 2005; Bolstad and Hipkins 2008; European Commission 2007). For example, in their meta-analysis, Minner et al. (2010) discuss the importance of student motivation and personal investment in the learning process as a key feature of inquiry-based learning. Inquiry-based science education (IBSE) means that students investigate their own questions to relevant scientific problems, gather and make sense of data and information, construct explanations, and convey their conclusions (Duschl et al. 2007; Lee et al. 2010). Such learning has been said to support understanding about the nature of science (NOS) and to contribute to lifelong learning practices (Bolstad and Hipkins 2008; Feldman et al. 2000; National Research Council 2012).

Within meaningful contexts, IBSE targets both the acquisition of process skills and discipline-based knowledge (Abd-El-Khalik et al. 2004). This means inquiry refers to both the process *and* a desired outcome, because students should not only attain skills such as learning how to ask questions but also learn about how knowledge in science is created, validated, and communicated. What is not always clear, however, is what counts as "real" and relevant in student investigations. For example, does authenticity in inquiry require that answers to questions are not known at all or does it suffice that

The original version of this chapter was revised. The erratum to this chapter is available at: DOI 10.1007/978-3-319-24615-4_7

students do not know them *yet* (Feldman et al. 2000)? If inquiry allows students to undertake their own investigations, they can take on more ownership over their learning and develop the skills of *how* to learn. Investigating authentic science problems also is alleged to change the nature of data collection by providing a purpose for activities that goes beyond validating what is already known. Not surprisingly, many curricula now make reference to inquiry and suggest teachers should support this learning approach by providing just enough scaffolding for students to accomplish tasks (Roth 2013).

One of the three dimensions of the United States' Next Generation Science Standards (NGSS) is *practices*. In *A Framework for K-12 Science Education*, the document that serves as the basis for the NGSS, *practices* represent the integration of skills and knowledge necessary for undertaking scientific inquiry (National Research Council 2012). As stated in *A Framework for K-12 Science Education*, "a focus on practices (in the plural) avoids the mistaken impression that there is one distinctive approach common to all science—a single 'scientific method'" (p. 48). Engaging students with practices is useful not only to promote interest and experience in various science disciplines but because it promotes a student understanding of science as a construction of knowledge that accumulates over time through a variety of methods. Furthermore, rather than promoting the idea that science inquiry is a linear process, *A Framework for K-12 Science Education* instead proposes a model of inquiry with three overlapping spheres of activity that scientists or engineers engage in as part of the process of inquiry: investigating, evaluating, and developing explanations and solutions.

As a second popular, constructivist approach to science education, problem-based learning (PBL) originated as an instructional strategy in medical schools, and now usage has extended to the K-12 and undergraduate setting (Savery and Duffy 1996). PBL is a learner-driven pedagogical approach. Students work in teams to propose methods or solutions to a real-world problem. In PBL, the teacher serves as a facilitator or guide and not as the primary source of knowledge. During PBL, students acquire new knowledge while also learning twenty-first-century skills such as collaboration and critical thinking. Much like IBSE, students employ diverse strategies to propose solution(s) to a given problem. Depending on the problem and design of a PBL unit, the overall process of inquiry may be constrained to a certain set of skills.

Several studies have highlighted the benefit of PBL. Strobel and van Barneveld (2009) performed a meta-synthesis of eight existing meta-analyses of PBL. Analysis of these large datasets indicated that PBL encouraged long-term retention, promoted skill development, and increased satisfaction of teachers and students (Strobel and van Barneveld 2009). Walker and Leary (2009) examined 82 studies of the effectiveness of PBL and determined that students engaged in PBL typically perform as well or better than their peers. An important caveat to PBL and other IBSE activities is that these approaches situate learning in complicated tasks and consequently require appropriate guidance and scaffolding to achieve success (Hmelo-Silver et al. 2007). We discuss this and other challenges of IBSE and PBL in the following section.

Challenges of IBSE and PBL

Despite the hoped-for learning benefits of inquiry-based science education (IBSE), mirroring authentic scientific inquiry in the classroom is difficult to achieve (Chinn and Malhotra 2002; Hakkarainen and Sintonen 2002; Scardamalia and Bereiter 1994). For both IBSE and PBL, there are challenges associated with teacher beliefs, knowledge, and lack of training as well as inconclusive relations between student activities and desired learning outcomes (Kirschner et al. 2006; Minner et al. 2010). A particular challenge of IBSE and PBL relevant to our current discussion is the need for a careful balance between providing students with support to conduct specific tasks while also nurturing independent thinking. For example, in IBSE, students position themselves ideally as experts when they present findings, but they are in the habit of falling back into a novice role when they receive feedback from others. To strengthen students' sense of expertise in science, support is needed for young people to confidently engage in reflective and critical discourse (Feldman et al. 2000). As we will discuss in the following section, support for such learner needs—advancing student understandings and skills—is best framed in terms of scaffolding.

References

Abd-El-Khalik, F., BouJaoude, S., Duschl, R., Lederman, N. G., Mamlok-Naaman, R., Hofstein, A., et al. (2004). Inquiry in science education: International perspectives. *Science Education, 88*(3), 397–419.

Aikenhead, G. S. (2005). *Science for everyday life: Evidence-based practice.* New York, NY: Teachers College Press.

Bolstad, R., & Hipkins, R. (2008). *Seeing yourself in science (Report).* Wellington, New Zealand: New Zealand Council for Educational Research.

Chinn, C. A., & Malhotra, B. A. (2002). Epistemologically authentic inquiry in schools: A theoretical framework for evaluating inquiry tasks. *Science Education, 86*(2), 175–218. doi:10.1002/sce.10001.

Duschl, R., Schweingruber, H., & Shouse, A. (Eds.). (2007). *Taking science to school: Learning and teaching science in grades K-8.* Washington, DC: National Academies Press.

European Commission. (2007). *Science education NOW: A renewed pedagogy for the future of Europe.* Community research report: Luxembourg: Office for Official Publications of the European Communities. Retrieved from internal-pdf://report-rocard-on-science-education_en--4063993856/report-rocard-on-science-education_en.pdf.

Feldman, A., Konold, C., Coulter, B., Conroy, B., Hutchison, C., & London, N. (2000). *Network science, a decade later: The internet and classroom learning.* Mahwah, NJ: Lawrence Erlbaum Associates.

Fensham, P. J. (2006). Humanistic science education: Moves from within and challenges from without. In *Proceedings of XII IOSTE symposium*, Penang, China.

Hakkarainen, K., & Sintonen, M. (2002). Interrogative model of inquiry and computer-supported collaborative learning. *Science & Education, 11*, 25–43.

Hmelo-Silver, C. E., Duncan, R. G., & Chinn, C. A. (2007). Scaffolding and achievement in problem-based and inquiry learning: A response to Kirschner, Sweller, and Clark (2006). *Educational Psychologist, 42*(2), 99–107. doi:10.1080/00461520701263368.

Kirschner, P. A., Sweller, J., & Clark, R. E. (2006). Why minimal guidance during instruction does not work: An analysis of the failure of constructivist, discovery, problem-based, experiential, and inquiry-based teaching. *Educational Psychologist, 41*(2), 75–86 [KO8].

Lee, H.-S., Linn, M. C., Varma, K., & Liu, O. L. (2010). How do technology-enhanced inquiry science units impact classroom learning? *Journal of Research in Science Teaching, 47*(1), 71–90. doi:10.1002/tea.20304.

Minner, D. D., Levy, A. J., & Century, J. (2010). Inquiry-based science instruction: What is it and does it matter? Results from a research synthesis years 1984 to 2002. *Journal of Research in Science Teaching, 47*(4), 474–496.

National Research Council. (2012). *A framework for K-12 science education: practices, crosscutting concepts, and core ideas.* Washington, DC: The National Academies Press.

OECD. (2006, May). Organisation for economic co-operation and development global science forum evolution of student interest in science and technology studies policy report.

Roth, W. M. (2013). *What more in/for science education: An ethnomethodological perspective.* Rotterdam: Sense Publishers.

Savery, J. R., & Duffy, T. M. (1996). Problem based learning: An instructional model and its constructivist framework. In B. G. Wilson (Ed.), *Constructivist learning environments: Case studies in instructional design* (pp. 135–150). Englewood Cliffs, NJ: Educational Technology Publications, Inc.

Scardamalia, M., & Bereiter, C. (1994). Computer support for knowledge-building communities. *The Journal of the Learning Sciences, 3*(3), 265–283.

Strobel, J., & van Barneveld, A. (2009). When is PBL more effective? A meta-synthesis of meta-analyses comparing PBL to conventional classrooms. *Interdisciplinary Journal of Problem-Based Learning, 3*(1). doi:10.7771/1541-5015.1046.

Walker, A., & Leary, H. (2009). A problem based learning meta analysis: Differences across problem types, implementation types, disciplines, and assessment levels. *Interdisciplinary Journal of Problem-Based Learning, 3*(1), 3–24. doi:10.7771/1541-5015.1061.

Chapter 5
Scaffolding Science Learning: Promoting Disciplinary Knowledge, Science Process Skills, and Epistemic Processes

Maggie Renken, Kathrin Otrel-Cass, Augusto Chioccariello, Isabelle Girault, and Melanie Peffer

In simple terms, "scaffolds" and "scaffolding" refer to support structures (scaffolds) that are provided (scaffolding) when novices or learners cannot work unassisted and require support to accomplish a task (Wood et al. 1976). In a formal learning situation, scaffolding ought to support students in achieving intended learning goals and tasks (Hmelo-Silver et al. 2007). The way scaffolds do this has been described by Pea (2004) based on Wood et al. (1976) and is achieved through channeling and focusing and through modeling. On one hand, channeling and focusing reduces freedom (channeling) and restricts the degree of choices (focusing) for a given task. This effectively simplifies the task. Modeling, on the other hand, allows for devising more advanced solutions to the given task through example.

Scaffolded instruction implies that the teacher is aware of students' needs for support regarding certain understanding or skills but goes further to provide the tools students need to accomplish a specific task at a certain time. As such, planning to use scaffolds forms part of the professional insight teachers have into their discipline, their students, and a particular task. This insight has been described as teachers' technological pedagogical content knowledge or TPACK (Mishra and Koehler 2006). Scaffolded instruction also requires teachers to be aware of how supporting scaffolds fall within the student's zone of proximal development (ZPD) (e.g., Lipscomb et al. 2012; Vygotsky 1978). Vygotsky and others describe ZPD as what a learner can do with or without assistance and emphasize that learners gradually develop the ability to accomplish certain tasks on their own. This process is often aided by the guidance and assistance of more knowledgeable others.

The original version of this chapter was revised. The erratum to this chapter is available at: DOI 10.1007/978-3-319-24615-4_7

M. Renken et al., *Simulations as Scaffolds in Science Education*,
SpringerBriefs in Educational Communications and Technology,
DOI 10.1007/978-3-319-24615-4_5

What Exactly Needs To Be Scaffolded in Science Education Contexts?

Since IBSE and PBL curricula can present particularly complex challenges for learners, scaffolds can assist in the process of developing more sophisticated understanding and skills to accomplish difficult tasks within a given, often constrained, time frame. To use scaffolds as part of instruction requires that teachers identify strategies that support the intended learning goals and reflect the needs of their students. In particular, it is important to consider the varied categories of student-centered knowledge and skill acquisition that are relevant to science education. Based on what has been proposed in the literature (c.f., de Jong and van Joolingen 1998; Reid et al. 2003) and by the National Research Council's Framework for K-12 Science Education (2012), we highlight domain knowledge, experimentation knowledge, and reflective processes as categories corresponding to three relevant layers of knowledge. We propose that in science education contexts, learning goals and student needs fall within these three categories and refer to them here as: discipline-based knowledge, experimentation or scientific process skills, and reflective and epistemic processes (Fig. 5.1).

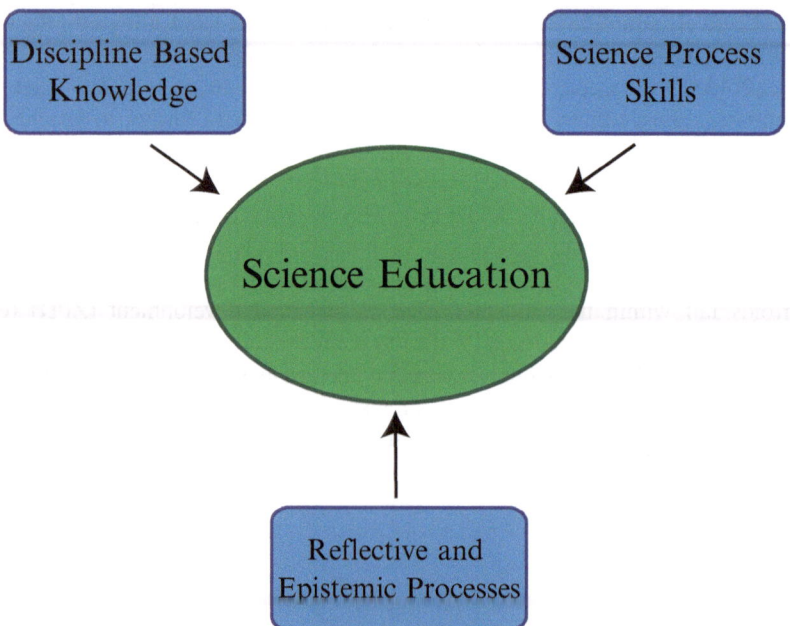

Fig. 5.1 Effective science education requires consideration of learning goals and student needs related to discipline-based knowledge, experimentation or scientific process skills, and reflective and epistemic processes

Discipline-Based Science Knowledge

School science has traditionally emphasized conceptual knowledge within given disciplines (e.g., biology, physics, chemistry, etc.). Evidence supports inquiry- and problem-based strategies for approaching conceptual understanding and, when necessary, conceptual change (Loyens et al. 2015; Minner et al. 2010). Students are also expected to develop skills in accessing key science concepts and interpreting or coordinating relevant information. Certain scaffolds may be especially useful for helping students make strides in learning how to access knowledge and construct their understanding. For example, some relevant scaffolds create access to appropriate sources of information, such as a website that contains descriptions of the studied concepts (Reid et al. 2003). Others make thinking visible—encouraging students to explain their understanding, in order to express the domain conceptual knowledge or to recognize their lack thereof (Linn and Bat-Sheva 2011).

Scientific Experimentation Knowledge and Skills

More recently, school science emphasizes the development of scientific process skills or practices (National Research Council 2012). Scientific reasoning literature suggests that many of these skills (e.g., hypothesis generation, control of variables) may not be easily developed without direct instruction or discovery-learning opportunities (see Zimmerman 2007 for a review of the development of scientific reasoning abilities). This calls for scaffolding learners as they systematically design scientific experiments, make predictions, observe outcomes, and draw reasonable conclusions. These scaffolds relate to the inquiry cycle as a whole (Linn and Bat-Sheva 2011; Puntambekar and Kolodner 2005).

Reflective and Epistemic Processes

There is evidence that students' self-awareness of the learning processes and their reflections, including the construction of abstractions and integration of discoveries with existing ideas, require understanding and abilities specific to the nature of knowledge and reflective processes. Relevant scaffolds then may concentrate on triggering or developing selected traits and abilities, such as curiosity or the ability to articulate ideas (Graesser et al. 2005; Quintana et al. 2004). For example, Linn et al. (2010) show that students' critique of simulated thermodynamics experiments is as effective as interacting with the simulation and more effective than observing the simulation.

Overall, IBSE and science PBL aim to equip students with general science practices and skills so they learn and experience what it means to think and work like a scientist. Developing such skills requires that students gain insights and learn about some of the tools scientists use to make sense of complex natural phenomena. This opens up the possibility for students to work with computer simulations, which have long been used by scientists to review, simplify, expand, and predict ideas and observations. Further, by requiring the use of increasingly more accessible information and communication technology (ICT) in the collection, analysis and representation of data in IBSE, students are likely to be engaged (Roth 2008). Science education, supported through ICT, may provide learners with opportunities for contextualized learning and "tinkering" with their developing ideas in science. Computer simulations may be a particularly useful form of ICT in this regard. As learning tools, simulations often constrain complexity to make ideas graspable without necessarily limiting opportunities for exploration. This makes simulations powerful, discovery-rich tools appropriate for incorporation in inquiry- and problem-based settings.

While it may seem at first contradictory to the nature of scientific inquiry that concepts are explicated and clarified but also constrained or limited through the use of simulations, such support structures are productive and not necessarily inauthentic. Just as scientists seek to isolate individual variables and constraints of a system to observe the effects of those specific variables, the design of an educational simulation might scaffold students' understanding by focusing their attention to relevant parameters (Finkelstein et al. 2005). Embedded within IBSE and PBL curricula rather than standing alone, the learning that simulations promote may not be limited to conceptual understanding. To date, however, empirical support for simulations as learning tools has focused primarily on conceptual understanding. There is moderate support for their impact on motivation and interest, but almost no evidence of their usefulness in promoting specific learning goals, such as those related to the nature of science (Honey and Hilton 2011; Kim et al. 2015).

Research on the benefits of activities that are well supported by simulations, like modeling, suggests simulations may indeed be valuable for promoting IBSE- and PBL-specific learning goals. For instance, a study considering the benefits of creating one's own models for learning and engaging in the process of metamodeling (i.e., understanding of the nature and purpose of scientific models) revealed improved student understanding of science content and nature of science understanding (Schwarz and White 2005). As we have described in Chap. 2 of this brief, simulations allow both scientists and students to use models to formulate new understanding of knowledge or as a method of communicating new findings (Schwarz et al. 2009). When considering the process of how students use models to attain understanding of science concepts, there are four elements to consider:

1. The construction of models based on prior evidence
2. The use of models to explain or predict phenomena
3. Evaluation of different models to represent or predict a phenomena
4. Revisions of models to improve their ability to explain or predict a phenomenon (Schwarz et al. 2009)

Involving learners in the construction of models instead of only in the evaluation of other models has several learning benefits. Since the generation of models is analogous to processes underlying human cognition, allowing students to engage in a process of creating, testing, and revising models about scientific content represents a concrete output of the internal cognitive process students are undertaking when learning the content (Schwarz and White 2005). By engaging in modeling, students are learning not only content but about the nature of science itself. For instance, if students understand science to be a process of iterative modeling building and refinement, it is easier for the students to understand all of human knowledge in terms of constructed knowledge that may or may not fully articulate certain phenomena (Schwarz and White 2005). A recent study of elementary age students found the majority of students equate knowledge building with task-based learning (Lunn Brownlee et al. 2015). The approach outlined here—to teaching students science with simulations embedded in inquiry- and problem-based curricula—is expected to alter such epistemic beliefs.

Although the benefits of engaging students in model-based inquiry are well supported, distinct challenges to implementing this approach remain. For example, teachers and students both often lack an understanding of what the purposes of models are or how they are used (Schwarz and White 2005). Effective and efficient scaffolding in inquiry-based learning is expected to be the result of synergistic teacher facilitation and lesson scaffolds (Hsu et al. 2015). Furthermore, student creation and modification of their own models is the least common form of modeling seen in schools (Schwarz et al. 2009). This is despite demonstrated benefits of students' modeling activities (see Namdar and Shen 2015 for review). In the following and final section, we explore considerations, such as lack of understanding of purpose, for educators, designers, and researchers to keep in mind when embedding simulations as scaffolds in the science classroom.

References

de Jong, T., & van Joolingen, W. R. (1998). Scientific discovery learning with computer simulations of conceptual domains. *Review of Educational Research, 68*(2), 179–201.

Finkelstein, N. D., Adams, W. K., Keller, C. J., Kohl, P. B., Perkins, K. K., Podolefsky, N. S., et al. (2005). When learning about the real world is better done virtually: A study of substituting computer simulations for laboratory equipment. *Physical Review Special Topics—Physics Education Research, 1*(1), 010103. doi:10.1103/PhysRevSTPER.1.010103.

Graesser, A. C., McNamara, D. S., & VanLehn, K. (2005). Scaffolding deep comprehension strategies through point & query, autotutor, and iSTART. *Educational Psychologist, 40*(4), 225–234.

Hmelo-Silver, C. E., Duncan, R. G., & Chinn, C. A. (2007). Scaffolding and achievement in problem-based and inquiry learning: A response to Kirschner, Sweller, and Clark (2006). *Educational Psychologist, 42*(2), 99–107. doi:10.1080/00461520701263368.

Honey, M., & Hilton, M. (Eds.). (2011). *Learning science through computer games and simulations*. Committee on Science Learning: Computer Games, Simulations, and Education. Board on Science Education, Division of Behavioral and Social Sciences and Education. Washington, DC: The National Academies Press.

Hsu, Y.-S., Lai, T.-L., & Hsu, W.-H. (2015). A design model of distributed scaffolding for inquiry-based learning. *Research in Science Education, 45*(2), 241–273. doi:10.1007/s11165-014-9421-2.

Kim, P., Suh, E., & Song, D. (2015). Development of a design-based learning curriculum through design-based research for a technology-enabled science classroom. *Educational Technology Research and Development, 63*(4), 575–602. doi:10.1007/s11423-015-9376-7.

Linn, M. C., & Bat-Sheva, E. (2011). *Science learning and instruction: Taking advantage of technology to promote knowledge integration—Routledge.* Retrieved from http://www.routledge.com/books/details/9780805860559/.

Linn, M. C., Chang, H.-Y., Chiu, J., Zhang, H., & McElhaney, K. (2010). Can desirable difficulties overcome deceptive clarity in scientific visualizations? In A. Benjamin (Ed.), *Successful remembering and successful forgetting: A Festschrift in honor of Robert A. Bjork* (pp. 239–262). New York: Routledge.

Lipscomb, L., Swanson, J., & West, A. (2012). *Scaffolding—Emerging perspectives on learning, teaching and technology.* Retrieved from http://projects.coe.uga.edu/epltt/index.php?title=Scaffolding.

Loyens, S. M. M., Jones, S. H., Mikkers, J., & van Gog, T. (2015). Problem-based learning as a facilitator of conceptual change. *Learning and Instruction, 38*, 34–42. doi:10.1016/j.learninstruc.2015.03.002.

Lunn Brownlee, J., Curtis, E., Spooner-Lane, R., & Feucht, F. (2015). Understanding children's epistemic beliefs in elementary education. *Education, 3–13*, 1–18. doi:10.1080/03004279.2015.1069369.

Minner, D. D., Levy, A. J., & Century, J. (2010). Inquiry-based science instruction—What is it and does it matter? Results from a research synthesis years 1984 to 2002. *Journal of Research in Science Teaching, 47*(4), 474–496. doi:10.1002/tea.20347.

Mishra, P., & Koehler, M. J. (2006). Technological pedagogical content knowledge: A framework for teacher knowledge. *Teachers College Record, 108*(6), 1017–1054.

Namdar, B., & Shen, J. (2015). Modeling-oriented assessment in K-12 science education: A synthesis of research from 1980 to 2013 and new directions. *International Journal of Science Education, 37*(7), 993–1023. doi:10.1080/09500693.2015.1012185.

National Research Council. (2012). *A framework for K-12 science education: Practices, crosscutting concepts, and core ideas.* Washington, DC: The National Academies Press.

Pea, R. D. (2004). The social and technological dimensions of scaffolding and related theoretical concepts for learning, education, and human activity. *The Journal of the Learning Sciences, 13*(3), 423–451.

Puntambekar, S., & Kolodner, J. L. (2005). Toward implementing distributed scaffolding: Helping students learn science from design. *Journal of Research in Science Teaching, 42*(2), 185–217. doi:10.1002/tea.20048.

Quintana, C., Reiser, B. J., Davis, E. A., Krajcik, J., Fretz, E., Duncan, R. G., et al. (2004). A scaffolding design framework for software to support science inquiry. *The Journal of the Learning Sciences, 13*(3), 337–386.

Reid, D. J., Zhang, J., & Chen, Q. (2003). Supporting scientific discovery learning in a simulation environment. *Journal of Computer Assisted Learning, 19*(1), 9–20. doi:10.1046/j.0266-4909.2003.00002.x.

Roth, W. M. (2008). Authentic science for all and the search for the ideal biology curriculum: A personal perspective. *Journal of Biology Education, 43*(1), 3–5.

Schwarz, C., Reiser, B., Davis, B., Kenyon, L., Acher, A., Fortus, D., et al. (2009). Designing a learning progression for scientific modeling: Making scientific modeling accessible and meaningful for learners. *Journal for Research in Science Teaching, 46*(6), 632–654.

Schwarz, C. V., & White, B. Y. (2005). Metamodeling knowledge: Developing students' understanding of scientific modeling. *Cognition and Instruction, 23*(2), 165–205. doi:10.1207/s1532690xci2302_1.

Vygotsky, L. S. (1978). *Mind and society: The development of higher psychological processes.* Cambridge, MA: Harvard University Press.

Wood, D., Bruner, J. S., & Ross, G. (1976). The role of tutoring in problem solving. *Journal of Child Psychology and Psychiatry and Allied Disciplines, 17*, 89–100.

Zimmerman, C. (2007). The development of scientific thinking skills in elementary and middle school. *Developmental Review, 27*(2), 172–223. doi:10.1016/j.dr.2006.12.001.

Chapter 6
Considerations for Integrating Simulations in the Science Classroom

Kathrin Otrel-Cass, Isabelle Girault, Maggie Renken, Augusto Chioccariello, and Melanie Peffer

In a prior review of 17 studies that investigated the use of simulations to supplement traditional instructions, the general conclusion was that including computer simulations as part of traditional instruction resulted in enhanced learning (Rutten et al. 2012). The two types of studies identified for review examined the use of simulations in physics and biology contexts (1) to improve student learning outcomes and (2) as a means to prepare students to engage in physical laboratory exercises. Specifically, computer simulations were found to be more time effective, to prepare students for participation in physical experiments, and to improve the students' overall satisfaction and perception of the science classroom. However, the reviewed studies only measured short-term gains, and the long-term benefits of including simulations as part of traditional instruction have yet to be assessed. Additionally, successful incorporation of simulations in the classroom requires specific considerations.

When teachers think about and plan to use simulations as scaffolds for their students' inquiry, they need to consider the *content*, *learner needs*, and *context needs* (van Joolingen et al. 2007). Recent trends in science education require teachers to consider conceptual knowledge and scientific reasoning skills as the necessary learning outcomes or *content* of science education. *Learner needs* include what a teacher knows specifically about an individual student and potentially diverse roadblocks in completing a task but also general, common problematic issues. For example, teachers may consider typical misconceptions for particular conceptual ideas as a departure point of their planning (e.g., see those discussed by Driver et al. 2013). Emphasis on *context needs* means that teachers have to consider the particularities of the settings within which they and their students are operating. This is particularly important for IBSE because it is the contextual nature that makes tools, such as simulations, potentially powerful to learners. To consider contextual factors,

The original version of this chapter was revised. The erratum to this chapter is available at: DOI 10.1007/978-3-319-24615-4_7

© AECT 2016

M. Renken et al., *Simulations as Scaffolds in Science Education*, SpringerBriefs in Educational Communications and Technology, DOI 10.1007/978-3-319-24615-4_6

teachers must think about technological aspects, including what can and cannot be done with the resources available. This requires particular attention to the a priori technological knowledge and skills with which students come equipped or still need assistance. Acknowledging the role of context also requires that teachers consider their own role and the role others (e.g., students or other learning community members) may play (van Joolingen et al. 2007).

Bringing all these complex aspects together means that teachers weave together instructional structures targeting and encompassing a range of support. Tabak (2004) describes this process as a synergy between different scaffold components and stresses the need to identify their characteristics. After all, scaffolds are meant not only to support *but* to integrate with a wide variety of learning activities (Dillenbourg et al. 2009). In fact, recall that simulations in the science classroom are best used when paired or otherwise combined with physical, or hands-on, experiments (de Jong et al. 2013). We recommend the following seven topics as considerations for successfully embedding simulations as scaffolds in science education:

Attending to Opportunities for Transfer during Curriculum Design

One challenge for teachers is to have students make use of the simplified ideas presented in simulations so they may then apply them to more authentic but complex systems. An approach for encouraging this transfer may be in the curriculum design itself. For example, by embedding a simplified simulation within a broader inquiry, students may begin to match models against physical experimental results. For example, a simulation of the hydrogen atom allows students to check how the prediction of various models, including Bohr's model, matches experimental results (Fig. 2.3).

Connecting Simulations Purposefully

Simulations should not serve as stand-alone exercises in the classroom. Instead, they should be meaningfully connected to the overall learning experience (Dillenbourg et al. 2009). This can be accomplished by purposing simulations to support learning goals *throughout* an inquiry. Simulations become less useful if they represent isolated phases of the inquiry activity and may be perceived by students to be unconnected or separate from their inquiries. Simulations should be integrated into IBSE curricula at particular points in a larger inquiry in order to provide a contiguous, more authentic, and more meaningful experience for students.

Developing Scientific Process Skills through the Manipulation of Variables

Simulations should be incorporated in IBSE to directly target students' Nature of Science knowledge or science practices. This is crucial not only for the educational task of addressing required learning outcomes but for expanding the currently limited research base considering the utility of simulations in achieving practice knowledge as well (Honey and Hilton 2011). Simulations are useful particularly because they offer unique opportunities to manipulate variables. Open-ended experiences manipulating variables allow students to acquire and consolidate scientific reasoning skills through direct instruction or discovery and expand and shape the knowledge they need for future inquiries—within simulated or real environments. The manipulation of variables in simulations allows students to not only test, but to observe and retest their hypotheses or questions, but to do so in an easily controlled and observed environment. Preparing students in such a way supports them working more independently over time.

Addressing Student Misconceptions

We mention above that simulations should focus on learning goals, but scaffolding also requires that students' learning needs are met. Oftentimes, at the heart of student needs are inconsistencies between their prior knowledge and the teachers' learning goals (Driver et al. 2013). Teachers need to uncover problematic areas, such as misconceptions; they can support and address with simulations before introducing such simulations. Student misconceptions should be addressed using simulations that provide guided experimental conditions and that challenge prior knowledge. For example, to understand the typically misconceived notion that mass does not affect the speed at which an object falls in a vacuum, students need to understand what it means to interact with objects of varying mass in a vacuous space. A simulated microworld may serve as an ideal scaffold here since such an experience cannot be offered in the classroom. It is important to note that this recommendation is weighted with the caveat that while simulations may provide unique opportunities for experimentation, it is not clear that this is enough for students to undergo conceptual change. Therefore, addressing student misconceptions with simulations must be an iterative process—incorporating simulations, perhaps cyclically, with other modes of instruction, such as hands-on experiments, reading texts, and classroom discussions requiring explanation and argumentation.

Fading Scaffolds to Best Support Learning

The complex integration of scaffolds in science education also requires appropriate fading of the scaffold as learning outcomes are achieved (Pea 2004). To use the context of difficult-to-observe phenomena as an example, consider the speed at which two objects of varying mass travel down the incline of a ramp. Because the crucial event for a student observing this phenomenon—the moment the objects reach the end of the ramp—may happen in an instance, experimentation may not result in a learner correctly acquiring discipline-based knowledge or greater conceptual understanding. For this example, it is likely that following a hands-on, physical experiment, the learner may rely on personal bias to draw some inaccurate conclusion or confirm his or her prior inaccurate conception (Renken and Nunez 2010). In this case, incorporating a simulation as a scaffold may aid not only in acquiring *correct* discipline knowledge but in developing reflective epistemic processes as well. For instance, the ability to stop, slow down, or precisely measure the speed of the two objects using a computer simulation may contribute to the learner's discipline knowledge but may also highlight for the learner the potential impact of his or her bias in drawing conclusions—an important metacognitive ability in authentic science practice. What is most important in this scenario is that the simulation scaffold should be removed once these disciplinary knowledge and reflective process learning goals are achieved in order for the learner to be engaged and to practice applying them in future, and potentially more complicated, situations.

Supporting Assessment for and of Learning

A significant aspect of incorporating simulations as scaffolds in science education relates to their role in both formative and summative assessment (De Klerk et al. 2015). So that both students and teachers have clear goals and outcomes in mind, assessment criteria must be clearly identified (van Joolingen et al. 2007). Clear goals and outcomes as well as assessment are important foundations for remaining apprised of learning, acknowledging where and when simulation scaffolds should be placed, and deciding when scaffolds should be faded or removed. While inquiry invites students to investigate their own questions, this does not mean that teachers cannot identify and assess specific learning outcomes. Even in an open inquiry, teachers set specific learning outcomes. With technological advances (e.g., educational data mining), simulations are increasingly useful with regard to formatively assessing these outcomes. This is not limited to teacher-based assessment. Instead, integrating simulations can play a significant part in the formative assessment process by allowing students, individually or in pairs or groups, to self assess their knowledge and abilities.

The successful integration of simulations as scaffolds in IBSE investigations and PBL activities should support formative and summative assessment strategies (c.f., Buckley and Quellmalz 2013). Formative assessment is especially important for identifying when scaffolds may be appropriately removed in order for learning to move forward. Simulations can provide fruitful opportunities for teachers to observe or engage in productive learning conversations with their students, when their students are working with what is on the screen. Dialogue that encourages students to talk about what they see, think about, and propose to do further supports and amplifies what the simulation has to offer and provides insight into students' ideas. It has been argued that scaffolded IBSE activities support students' evaluating their growing competencies both in gaining understanding and applying this knowledge and by reducing cognitive load and opening opportunities for dealing with more complex domains as well as collaboration and self-directed learning (Hmelo-Silver, Duncan & Chinn 2007).

Building Teacher Knowledge

If simulations are to provide scaffolds for students' conceptual understanding and for future investigations, teachers need to prepare for this. Selecting appropriate simulations at the right time as part of a scaffolding strategy to address specific learning needs requires that teachers build on and develop their technological pedagogical content knowledge (TPACK) (Mishra and Koehler 2006). Their situated insight into what their learners know already and may be struggling with, combined with their understanding of the learning opportunities simulations, in addition to other instructional activities, offer will help them to identify needs and strategies to support student inquiries through simulations. To foster such competencies, teacher pre- or in-service courses should include computer simulations as teaching resources. This will give teachers opportunities to explore how and when to support (or not support) student learning through computer simulations and to further consider how this supports authentic school science investigations.

Participating in inquiry-based science education or problem-based learning does not mean that students have to venture unaided into unknown problems. The key to successful inquiries is that teachers equip their students with the necessary skills so they can solve problems of growing complexity. Computer simulations have the potential to support the various stages students go through in an inquiry and do not necessarily diminish its authenticity. Most importantly, although prior research has emphasized the role of simulations in learners' conceptual understanding, we suggest when simulations are employed as scaffolds within a larger inquiry, they have the potential to impact not only discipline-based science knowledge but scientific experimentation practices and skills and reflective and epistemic practices as well. The successful application will depend on the development, implementation, and empirical consideration of simulations embedded within IBSE as scaffolds that are based on content, learner needs, and context needs.

References

Buckley, B. C., & Quellmalz, E. S. (2013). Supporting and assessing complex biology learning with computer-based simulations and representations. In D. F. Treagust & C.-Y. Tsui (Eds.), *Multiple representations in biological education* (pp. 247–267). Netherlands: Springer. Retrieved from http://link.springer.com/chapter/10.1007/978-94-007-4192-8_14.

de Jong, T., Linn, M. C., & Zacharia, Z. C. (2013). Physical and virtual laboratories in science and engineering education. *Science, 340*(6130), 305–308. doi:10.1126/science.1230579.

De Klerk, S., Veldkamp, B. P., & Eggen, T. J. H. M. (2015). Psychometric analysis of the performance data of simulation-based assessment: A systematic review and a Bayesian network example. *Computers & Education, 85*, 23–34. doi:10.1016/j.compedu.2014.12.020.

Dillenbourg, P., Järvelä, S., & Fischer, F. (2009). The evolution of research on computer-supported collaborative learning. In N. Balacheff, S. Ludvigsen, T. De Jong, A. Lazonder, S. Barnes, & L. Montandon (Eds.), *Technology-enhanced learning* (pp. 3–19). Netherlands: Springer. Retrieved from http://link.springer.com/chapter/10.1007/978-1-4020-9827-7_1.

Driver, R., Rushworth, P., Squires, A., & Wood-Robinson, V. (2013). *Making sense of secondary science: Research into children's ideas*. London: Routledge.

Hmelo-Silver, C. E., Duncan, R. G., & Chinn, C. A. (2007). Scaffolding and Achievement in Problem-Based and Inquiry Learning: A Response to Kirschner, Sweller, and Clark (2006). *Educational Psychologist, 42*(2), 99–107. doi:10.1080/00461520701263368.

Honey, M., & Hilton, M. (Eds.). (2011). *Learning science through computer games and simulations*. Committee on Science Learning: Computer Games, Simulations, and Education. Board on Science Education, Division of Behavioral and Social Sciences and Education. Washington, DC: The National Academies Press.

Mishra, P., & Koehler, M. J. (2006). Technological pedagogical content knowledge: A framework for teacher knowledge. *Teachers College Record, 108*(6), 1017–1054.

Pea, R. D. (2004). The social and technological dimensions of scaffolding and related theoretical concepts for learning, education, and human activity. *The Journal of the Learning Sciences, 13*(3), 423–451.

Renken, M. D., & Nunez, N. (2010). Evidence for improved conclusion accuracy after reading about rather than conducting a belief-inconsistent simple physics experiment. *Applied Cognitive Psychology, 24*(6), 792–811. doi:10.1002/acp.1587.

Rutten, N., Van Joolingen, W. R., & Van Der Veen, J. T. (2012). The learning effects of computer simulations in science education. *Computers and Education, 58*(1), 136–153. doi:10.1016/j.compedu.2011.07.017.

Tabak, I. (2004). Synergy: A complement to emerging patterns of distributed scaffolding. *The Journal of the Learning Sciences, 13*(3), 305–335.

van Joolingen, W. R., De Jong, T., & Dimitrakopoulou, A. (2007). Issues in computer supported inquiry learning in science. *Journal of Computer Assisted Learning, 23*(2), 111–119. doi:10.1111/j.1365-2729.2006.00216.x.

Simulations as Scaffolds in Science Education

**Maggie Renken, Melanie Peffer, Kathrin Otrel-Cass,
Isabelle Girault, and Augusto Chiocarriello**

© AECT 2016
M. Renken et al., *Simulations as Scaffolds in Science Education*,
SpringerBriefs in Educational Communications and Technology,
DOI 10.1007/978-3-319-24615-4

DOI 10.1007/978-3-319-24615-4_7

The publisher regrets that the affiliation of the series editor M.J. Bishop on page i is incorrect in the print and online versions of this book. The correct information is given below.

University System of Maryland in Adelphi, MD

The online version of the updated book can be found at
http://dx.doi.org/10.1007/978-3-319-24615-4

© AECT 2016
M. Renken et al., *Simulations as Scaffolds in Science Education*,
SpringerBriefs in Educational Communications and Technology,
DOI 10.1007/978-3-319-24615-4_7

The names of the authors' for each chapter below were omitted at the time of publication. Updated now are the correct order in which their names should appear.

Chapter 1
An Introduction to Simulations as Scaffolds in Science Education

Maggie Renken, Kathrin Otrel-Cass, Melanie Peffer, Isabelle Girault, and Augusto Chioccariello

The online version of the updated chapter can be found at http://dx.doi.org/10.1007/978-3-319-24615-4_1

Chapter 2
Computer Simulations on a Multidimensional Continuum: A Definition and Examples

Isabelle Girault, Melanie Peffer, Augusto Chioccariello, Maggie Renken, and Kathrin Otrel-Cass

The online version of the updated chapter can be found at http://dx.doi.org/10.1007/978-3-319-24615-4_2

Chapter 3
Distinctions Between Computer Simulations and Other Technologies for Science Education

Melanie Peffer, Maggie Renken, Isabelle Girault, Augusto Chioccariello, and Kathrin Otrel-Cass

The online version of the updated chapter can be found at http://dx.doi.org/10.1007/978-3-319-24615-4_3

Chapter 4
Inquiry-Based Science Education and Problem-Based Learning: Motivations, Objectives, and Challenges Relevant to Computer Simulations

Kathrin Otrel-Cass, Maggie Renken, Melanie Peffer, Isabelle Girault, and Augusto Chioccariello

The online version of the updated chapter can be found at http://dx.doi.org/10.1007/978-3-319-24615-4_4

Chapter 5
Scaffolding Science Learning: Promoting Disciplinary Knowledge,
Science Process Skills, and Epistemic Processes

Maggie Renken, Kathrin Otrel-Cass, Augusto Chioccariello,
Isabelle Girault, and Melanie Peffer

The online version of the updated chapter can be found at
http://dx.doi.org/10.1007/978-3-319-24615-4_5

Chapter 6
Considerations for Integrating Simulations in the Science Classroom

Kathrin Otrel-Cass, Isabelle Girault, Maggie Renken,
Augusto Chioccariello, and Melanie Peffer

The online version of the updated chapter can be found at
http://dx.doi.org/10.1007/978-3-319-24615-4_6

About the Authors

Maggie Renken, Ph.D.

Maggie is an Assistant Professor in the Educational Psychology program at Georgia State University. Broadly, her research focuses on scientific thinking and the acquisition of science knowledge. Her work has considered how adolescents and young adults learn and alter inaccurate prior beliefs through various media, including computer simulations, text, and hands-on experimentation. This research is intended to inform approaches for assessing and improving reasoning and thinking skills.

Melanie Peffer, Ph.D.

Melanie has a Ph.D. in molecular biology and is a postdoctoral associate in the Educational Psychology department at Georgia State University. Melanie's research program integrates her training in molecular biology and the learning sciences to create a synergistic program of study. Her work examines student learning during authentic inquiry.

Kathrin Otrel-Cass, Ph.D.

Kathrin is an Associate Professor in Science Education. She leads the techno-anthropology video lab and is the co-leader of the techno-anthropology research group. Kathrin is interested in ICT and its role in mediating learning and practices in science and technology and in science education. Kathrin also is interested in culturally responsive pedagogy and the nature of interactions in science and

© AECT 2016   35
M. Renken et al., *Simulations as Scaffolds in Science Education*,
SpringerBriefs in Educational Communications and Technology,
DOI 10.1007/978-3-319-24615-4

technology. Kathrin is an experienced classroom-based researcher and has published on assessment for learning, culture in science education, and classroom-based research methods. She has lived and researched internationally in Austria and New Zealand before coming to Denmark.

Isabelle Girault, Ph.D.

Isabelle is an Associate Professor at the University of Grenoble-Alpes (France). Her research concerns Chemistry Education and learning sciences through a scientific inquiry process with the use of ICT. This research focuses on how a computer environment can scaffold the activity of experimental design and more precisely the elaboration of protocols by students through automatic feedbacks.

Augusto Chiocarriello, Ph.D.

Augusto is a Researcher at the Italian National Research Council's Institute for Educational Technology. He obtained his Physics degree (magna cum laude) in 1980 at the University of Naples. From 1982 to 1986, he worked in physics education at the Educational Technology Center, UC Irvine, initially as a National Research Council research fellow and subsequently as project manager. In 1986, he joined the Institute for Educational Technology as a researcher, and he worked on exploiting multimedia technology in design and development of learning systems in several EC DELTA projects. Since 1995, he has collaborated with Reggio Emilia infant schools, exploring the use of computational construction kits as learning tools for early childhood education. More recently, he has coordinated the institute's participation in the IST-WebLabs project.

Index

© AECT 2016
M. Renken et al., *Simulations as Scaffolds in Science Education*,
SpringerBriefs in Educational Communications and Technology,
DOI 10.1007/978-3-319-24615-4